KV-511-204

This workbook has been written and developed to be used alongside the Lonsdale revision guide, AQA GCSE Science Essentials, to help you get the most out of your revision. You can use it if you are studying AQA GCSE Science A or B at Foundation or Higher Tier.

It contains 'quick fire' questions, including multiple choice questions, matching pair exercises (like those in the Science A objective tests) and short answer questions, to test your understanding of the topics covered in the revision guide.

Start by reading through a topic in the revision guide. Make notes, jotting down anything that you think will help you to remember the information. When you have finished, take a short break and then read through your notes. You might even want to try covering them up and writing them out again from memory.

Finally, work through the relevant questions in this workbook without looking at the guide or your notes.

The page headers and sub-headings in this workbook correspond with those in your revision guide, so that you can easily identify the questions that relate to each topic.

Completing the questions will help to reinforce your understanding of the topics covered in the guide and highlight any areas that need further revision.

The answers to the questions in this workbook are available in a separate booklet. There is a box at the end of each page for you to record your score. Don't worry if you get some questions wrong the first time. Just re-read the information in your revision guide and try again.

The tick boxes on the contents page let you track your revision progress; just put a tick in the box next to each topic when you are confident that you understand it.

Good luck with your exams!

The Nervous System

The Parts of the Nervous System

1 a) What does the central nervous system (CNS) consist of? Tick the correct option.

A Brain and effectors ☐   B Brain and spinal cord ☐

C Spinal cord and receptors ☐   D Receptors and effectors ☐

b) Which structure is a sensory receptor? Tick the correct option.

A The liver ☐   B The kidney ☐

C The skin ☐   D The stomach ☐

c) What is another name given to nerve cells? Tick the correct option.

A Capillaries ☐   B Effectors ☐

C Neurones ☐   D Receptors ☐

The Three Types of Neurone

2 Fill in the missing words to complete the following sentences.

The _____ neurones receive messages from the receptors and send them to the CNS.

The _____ neurones send messages from the CNS to the receptors.

These two neurones are connected in the CNS by a _____ neurone.

3 An electrical signal only travels in one direction down a neurone.

Is this statement true or false? _____

Connections Between Neurones

4 a) What is the name of the gap between two nerve cells?

_____

b) What happens when an electrical, or nervous, impulse reaches the gap between two neurones? Tick the correct option.

A The impulse stops ☐

B The impulse jumps the gap ☐

C A chemical transmitter is released ☐

D A hormone is released ☐

Types of Receptor

1 List three stimuli and the receptors used to detect them.

a) _____

b) _____

c) _____

2 a) Which of the following shows the correct sequence of events in the passage of a nerve impulse? Tick the correct option.

A Effector → Receptor → Sensory neurone → CNS → Motor neurone ☐

B Receptor → Sensory neurone → CNS → Motor neurone → Effector ☐

C Receptor → Motor neurone → CNS → Sensory neurone → Effector ☐

D Effector → Sensory neurone → CNS → Motor neurone → Receptor ☐

b) Which part of the nervous system acts as the coordinator in the passage of a nerve impulse? Tick the correct option.

A Effector ☐   B Receptor ☐

C Brain ☐   D Central Nervous System (CNS) ☐

Reflex Action

3 a) Fill in the missing words to complete the following sentences.

In a _____ arc the electrical impulse does not enter the conscious areas of your brain. The time between the stimulus and the _____ is as short as possible.

The only three neurones involved in a reflex arc are the sensory neurones, the _____ neurones and the _____ neurones.

b) Sometimes conscious action is too slow to prevent harm to the body. A reflex action speeds up the response time by missing out the brain completely.

Which of the following is an example of a reflex action? Tick the correct option.

A Looking both ways to cross the road ☐   B Removing your hand from a hot plate ☐

C Laughing at a joke ☐   D Whistling to get someone's attention ☐

Contents

Contents

The Nervous System

The Parts of the Nervous System

1 **a)** What does the central nervous system (CNS) consist of? Tick the correct option.

    A Brain and effectors ☐   **B** Brain and spinal cord ☑

    C Spinal cord and receptors ☐   **D** Receptors and effectors ☐

b) Which structure is a sensory receptor? Tick the correct option.

    A The liver ☐   **B** The kidney ☐

    C The skin ☑   **D** The stomach ☐

c) What is another name given to nerve cells? Tick the correct option.

    A Capillaries ☐   **B** Effectors ☐

    C Neurones ☑   **D** Receptors ☐

The Three Types of Neurone

2 Fill in the missing words to complete the following sentences.

The _____sensory_____ neurones receive messages from the receptors and send them to the CNS.

The _____motor_____ neurones send messages from the CNS to the effectors.

These two neurones are connected in the CNS by a _____relay_____ neurone.

3 An electrical signal only travels in one direction down a neurone.

Is this statement true or false? _____True_____

Connections Between Neurones

4 **a)** What is the name of the gap between two nerve cells?

_____The synapse_____

b) What happens when an electrical, or nervous, impulse reaches the gap between two neurones? Tick the correct option.

    A The impulse stops ☐

    B The impulse jumps the gap ☐

    C A chemical transmitter is released ☑

    D A hormone is released ☐

The Nervous System

Types of Receptor

1 List three stimuli and the receptors used to detect them.

a) Light is detected by the receptors in the eyes

b) Sound is detected by the receptors in the ears

c) Taste is detected by the receptors in the tongue

2 a) Which of the following shows the correct sequence of events in the passage of a nerve impulse? Tick the correct option.

A Effector → Receptor → Sensory neurone → CNS → Motor neurone ☐

B Receptor → Sensory neurone → CNS → Motor neurone → Effector ☑

C Receptor → Motor neurone → CNS → Sensory neurone → Effector ☐

D Effector → Sensory neurone → CNS → Motor neurone → Receptor ☐

b) Which part of the nervous system acts as the coordinator in the passage of a nerve impulse? Tick the correct option.

A Effector ☐   B Receptor ☐

C Brain ☐   D Central Nervous System (CNS) ☑

Reflex Action

3 a) Fill in the missing words to complete the following sentences.

In a ___Reflex___ action the electrical impulse does not enter the conscious areas of

your brain. The time between the stimulus and the ___effector___ is as short as possible.

The only three neurones involved in a reflex action are the sensory neurones, the

___relay___ neurones and the ___motor___ neurones.

b) Sometimes conscious action is too slow to prevent harm to the body. A reflex action speeds up the response time by missing out the brain completely.

Which of the following is an example of a reflex action? Tick the correct option.

A Looking both ways to cross the road ☐   B Removing your hand from a hot plate ☑

C Laughing at a joke ☐   D Whistling to get someone's attention ☐

Internal Environment and Hormones

Internal Conditions

1 **a)** Fill in the missing words to complete the following sentences.

Humans need to keep their internal _____environment_____ relatively _____constant_____.

This is done by _____ things such as temperature and water content.

b) The amount of water in the body must be controlled. Suggest three ways in which water is removed from the body.

i) _____Water is removed by sweat_____

ii) _____Water is removed by urine_____

iii) _____Water is removed by lungs when you breathe_____

c) How does the body take in ions? Tick the correct option.

A By sweating ⬭   **B** By breathing ⬭

C By eating and drinking ☑   **D** By excretion ⬭

d) How is glucose used up? Tick the correct option.

A As energy in movement ☑   **B** In keeping cool ⬭

C In digestion ⬭   **D** In sweating ⬭

Hormones

2 **a)** Where are hormones produced?

_____Hormones are produced by glands (pituitary gland & ovaries)_____

b) How are hormones transported around the body? Tick the correct option.

A By the nervous system ⬭   **B** In the air we breathe in ⬭

C By the skin ⬭   **D** In the bloodstream ☑

3 **a)** Which hormone is produced by the pituitary gland? Tick the correct option.

A Adrenalin ⬭   **B** FSH ☑

C Oestrogen ⬭   **D** Progesterone ⬭

b) Which hormone is produced by the ovaries? Tick the correct option.

A Adrenalin ⬭   **B** FSH ⬭

C Oestrogen ☑   **D** LH ⬭

Internal Environment and Hormones

Natural Control of Fertility

1 a) The diagram represents the female reproductive system. Label the main parts.

A _Ovary_

B _Uterus_

b) Draw lines between the boxes to match each hormone to its function.

Hormone	Function
① FSH	③ Stimulates the release of an egg
② Oestrogen	① Causes the ovaries to produce oestrogen
③ LH	② Slows down the production of FSH

Artificial Control of Fertility

1 a) Which of the following is a treatment for infertile couples? Tick the correct option.

A Contraception ◯   B Ovulation ◯

C IVF ◯   D LH ☑

b) Which of the following can be used as an oral contraceptive? Tick the correct option.

A Progesterone ◯   B FSH ☑

C LH ◯   D Oestrogen ☑

c) i) How is FSH used in the artificial control of fertility?

FSH is given to a woman to increase fertility (as a drug) if a woman doesn't produce FSH naturally to stimulate eggs to mature.

ii) How is oestrogen used in the artificial control of fertility?

Oestrogen reduces fertility as an oral contraceptive to slow down FSH production so that eggs don't mature

Diet

Metabolic Rate

1 What does the term **metabolic rate** refer to? Tick the correct option.

A How quickly the heart beats ☐

B How quickly food is digested ☐

C How quickly chemical reactions in our body cells take place ☑

D How quickly we exercise ☐

2 Circle the correct options in the following sentences.

Your metabolic rate stays **low** / **high** for some time after **exercise** / **sleep.** The more exercise you take, the fitter you are likely to be. The more exercise you take, the **more** / **less** food you need.

Healthy Diets

3 a) A combination of what two things are needed to keep the body healthy.

i)   Excersize

ii)   healthy diet

b) Which of the following diseases might a person suffer from if they are overweight?
Tick the **four** correct options.

A Arthritis ☑   **B** Diabetes ☑

C Cystic fibrosis ☐   **D** Anorexia nervosa ☐

E Low blood pressure ☐   **F** Sickle cell anaemia ☐

G Heart disease ☑   **H** Stroke ☑

c) What is the name given to diseases caused by a lack of vitamins and minerals?
Tick the correct option.

A Inadequacy diseases ☐   **B** Insufficiency diseases ☐

C Diet diseases ☐   **D** Deficiency diseases ☑

d) What is the cause of malnourishment? Tick the correct option.

A Not enough vitamins in the diet ☑   **B** Not eating enough food ☐

C Not eating a balanced diet ☐   **D** Not enough protein in the diet ☐

e) Are you more likely to be obese if you live in the **developed** world or the **developing** world?

You are more likely to be obese in the developed world

f) Are you more likely to have problems due to the lack of food in the **developed** world or the

developing world? You are more likely to have lack of food in the developing world. ☐

Cholesterol

1 a) Where is cholesterol made in the body? Tick the correct option.

A The heart ◯   B The liver ✓

C The kidneys ◯   D The stomach ◯

b) Circle the correct options in the following sentences.

Cholesterol is found in the **blood** / **kidneys**. The level of cholesterol is influenced by **diet** / **exercise** and inherited factors. Cholesterol is carried in the blood by **hormones** / **lipoproteins**.

2 a) What does LDL stand for?

LDL stands for Low density lipoproteins

b) What does HDL stand for?

HDL stands for High density lipoproteins

c) Which of the following is classed as 'bad' cholesterol? Tick the correct option.

A HDL ◯   B LDL ✓

C GDL ◯   D BDL ◯

Fats and Salts

3 Which one of the following types of fats increases blood cholesterol and blood sugar levels? Tick the correct option.

A Saturated fats ✓   B Monounsaturated fats ◯

C Polyunsaturated fats ◯   D Vegetable oil ◯

4 What can happen if a person's diet contains too much salt? Tick the correct option.

A Blood pressure goes down ◯   B Cholesterol is reduced ◯

C Blood pressure goes up ✓   D Mineral ions levels are reduced ◯

5 Why is it better to prepare your own meals from raw ingredients than to buy processed food and ready meals?

It would be more healthy preparing your own meals from raw ingredients because processed food and ready meals have a higher intake of salt and saturated fats.

Drugs

Drugs

1 Which two types of source can drugs be made from?

a) Drugs can be made from scientists

b) Drugs can be made from natural substances

Developing New Drugs

2 a) Fill in the missing words to complete the sentences below.

New medical drugs must be ___Tested___ and ___Trialed___ to find out

whether they are ___toxic___.

They must then be checked for ___Side affects___.

b) What does **toxic** mean?

Toxic means something is poisonous

Thalidomide

3 a) Which condition was thalidomide first developed to treat? Tick the correct option.

A Insomnia ✓   B Depression ☐

C Obesity ☐   D Heart disease ☐

b) Why did pregnant women start to take thalidomide? Tick the correct option.

A It prevented miscarriage. ☐   B It stopped morning sickness. ✓

C It stopped multiple births. ☐   D It prevented ovulation. ☐

c) What were the effects of taking thalidomide during pregnancy? Tick the correct option.

A Women gave birth prematurely. ☐

B Babies were born with severe limb abnormalities. ✓

C Women had multiple births. ☐

D Babies had a high birth weight. ☐

d) Which disease is thalidomide now used to treat?

Thalidomide now treats Leprosy.

Drugs

Legal and Illegal Drugs

1 Fill in the missing words to complete the sentences below.

Illegal drugs, such as heroin and _cocaine_, and legal drugs, such as

alcohol and tobacco are used by some people for _recreation_. But they

can be very _addictive_.

Drugs alter chemical processes in the body so people can become _dependent_ to them.

2 Which of the following drugs are legal? Tick the **three** correct options.

A Heroin ☐   B Ecstasy ☐

C Alcohol ✓   D Cannabis ☐

E Nicotine ✓   F Caffeine ✓

Alcohol and Tobacco

3 a) What is the addictive substance found in tobacco? Tick the correct option.

A Tar ☐   B Cannabis ☐

C Ash ☐   D Nicotine ✓

b) Which gas, found in cigarette smoke, prevents red blood cells from carrying oxygen? Tick the correct option.

A Oxygen ☐   B Carbon dioxide ☐

C Carbon monoxide ✓   D Nitrogen ☐

c) Smoking during pregnancy can deprive the developing fetus of oxygen. How might this affect the birth mass of the baby?

The body wouldn't recieve enough oxygen through the red blood cells

d) Smoking can cause a number of harmful diseases. Name two diseases caused by smoking cigarettes.

i) _Lung cancer_   ii) _bronchitis_

4 a) Which two organs are affected by drinking too much alcohol? Tick the correct option.

A Heart and stomach ☐   B Brain and liver ✓

C Kidneys and liver ☐   D Brain and heart ☐

b) Why are car accidents more likely to happen if the driver is under the influence of alcohol?

because alcohol slows down reactions

Pathogens

1 What are pathogens?

Pathogens are microorganisms that cause infectious diseases

2 Circle the correct words in the following sentences.

Viruses are **smaller** / **bigger** than bacteria. They reproduce **slowly** / **quickly** inside living cells. They produce **toxins** / **antitoxins**, and cause illnesses.

3 a) Which of the following is a disease caused by a bacterium? Tick the correct option.

A Ringworm ⬜

B Food poisoning ⬜

C Polio ⬜

D Tuberculosis ☑

E Measles ⬜

F Cancer ⬜

b) Which of the following is a disease caused by a virus? Tick the correct option.

A Tetanus ⬜

B Flu ☑

C Athlete's foot ⬜

D Ringworm ⬜

E Cholera ⬜

F Cancer ⬜

4 Which part of our blood defends the body against infection? Tick the correct options.

A Red blood cells ⬜

B Plasma ⬜

C Platelets ⬜

D White blood cells ☑

5 a) What do antitoxins do? Tick the **two** correct options.

A They destroy the toxins produced by viruses. ☑

B They destroy the antibodies produced by bacteria. ⬜

C They destroy the toxins produced by bacteria. ⬜

D They destroy the antigens produced by viruses. ⬜

b) What do antibodies do? Tick the correct option.

A They destroy the toxins produced by pathogens. ⬜

B They attack healthy body cells. ⬜

C They destroy certain pathogens. ☑

D They ingest pathogens. ⬜

Treatment of Disease

Revision Guide Reference: Page 19

1 Which of the following is an example of a painkiller? Tick the correct option.

A Alcohol ☐   **B** Aspirin ☑

C Insulin ☐   **D** Vitamin ☐

2 Fill in the missing words to complete the sentences below.

The symptoms of a disease can often be alleviated using _painkillers_. However, these drugs

do not _kill_ pathogens. Antibiotics can be used to kill infective

bacterial pathogens. But they cannot kill _viral_ pathogens, which

live and _reproduce_ inside the body's cells.

3 **a)** Give an example of a strain of bacteria which has developed resistance to antibiotics.

A bacteria which has developed resistance is MRSA

b) By what process do bacteria develop resistance to antibiotics?

The bacteria developed resistance by the process of natural selection

Vaccination

4 **a)** What do vaccines contain? Tick the correct option.

A Medical drugs ☐   **B** Antibiotics ☐

C Dead / harmless forms of the disease ☑   **D** Antibodies ☐

b) Write **true** or **false** alongside each of the following statements about vaccines.

i) Red blood cells produce antibodies. _false_

ii) Antibiotics destroy the antibodies. _false_

iii) A live pathogen is injected into your body. _false_

iv) Vaccines provide an acquired immunity. _true_

c) What three illnesses does the MMR vaccine protect against?

i) _Measles_   **ii)** _Mumps_

iii) _Rubella_

Competition and Adaptations

Population and Communities

1 Which of the following statements is the correct definition of the word **population**? Tick the correct option.

    A The total number of organisms living in a particular habitat. ☐

    B The total number of individuals of the same species living in a particular habitat. ☑

    C The total number of animals living in a particular habitat. ☐

    D The total number of plants living in a particular habitat. ☐

2 Which of the following statements describes a **community**? Tick the correct option.

    A A group of animals and plants interacting with one another. ☑

    B A population of animals. ☐

    C Animals adapted to their surroundings. ☐

    D A food chain. ☐

Competition

3 Which of the following options do animals need to compete for? Tick the **three** correct options.

    A Light ☐   **B** Food ☑

    C Water ☑   **D** Space ☑

4 What one thing do plants compete for that animals do not? Tick the correct option.

    A Light ☑   **B** Nutrients ☑

    C Space ☐   **D** Water ☑

5 a) Fill in the missing words to complete the sentences.

When organisms compete, those which are _____ **better** _____ **adapted** _____ to their environment are more _____ **successful** _____ and usually exist in _____ **large** _____ numbers.

b) What would you expect to happen to the population size of an organism that was less well adapted to its environment than another?

They would not recieve their right things for survival and would die out.

Competition and Adaptations

Adaptations

1 a) Which of the following statements best describes what the word **adaptation** means? Tick the correct option.

A A significant change in the size of a population. ◯

B A feature that develops to make an organism better suited to its environment. ☑

C The process by which the different organisms in a habitat learn to share resources. ◯

D A characteristic that is acquired by an individual, e.g. a scar. ◯

b) What is the advantage of an adaptation?

The animal or plant would be more likely to survive

2 a) Briefly describe the conditions in which a polar bear lives.

The polar bear lives in cold conditions where there is snow and ice (so they have developed fur and a thick layer of fat).

b) Which of the following are adaptations of the polar bear? Tick the correct option(s).

A Large ears ◯

B Thick coat ☑

C Layer of fat under skin ☑

D Small feet ◯

E Camouflage ☑

F Large surface area to volume ratio ☑

3 Cacti have spines instead of leaves. How do these spines help them to survive? Tick the **two** correct options.

A They help to trap and store rain water. ◯

B They reflect the Sun's heat. ◯

C They minimise surface area to reduce water loss. ◯

D They make the plant look like a predator. ☑

E They prevent animals from eating the plant. ☑

F They trap insects, which the plant feeds on. ◯

Genetics

Genetic Information

1 What are the sections of chromosome called that control our characteristics? Tick the correct option.

A Genes ☑   **B** DNA ☐

C Helix ☐   **D** Amino acids ☐

2 How many pairs of chromosomes do humans have?

23 chromosomes

Variation

3 Which of the following characteristics are due to genetic variation? Tick the correct options.

A Eye colour ☑   **B** Religion ☐

C Tongue rolling ability ☐   **D** Weight ☐

4 Which of the following characteristics are due to environmental variation? Tick the correct options.

A Accent ☑   **B** Eye colour ☐

C Weight ☑   **D** Scars ☑

Reproduction and Variation

5 Which type of reproduction involves only one parent? Tick the correct option.

A Asexual reproduction ☑   **B** Sexual reproduction ☐

C Evolution ☐   **D** Natural selection ☐

6 a) Which type of reproduction leads to variation? Tick the correct option.

A Evolution ☐   **B** Sexual reproduction ☑

C Natural selection ☐   **D** Asexual reproduction ☐

b) Briefly explain why there is variation in the offspring of this type of reproduction.

7 What are clones? Tick the correct option.

A Genetically identical individuals ☑   **B** Genetically different individuals ☐

C Individuals formed by sexual reproduction ☐   **D** Individuals that show variation ☐

Reproducing Plants

1 By which method do plants reproduce naturally? Tick the correct option.

   A Sexually ☐   **B** Genetic modification ☐

   C Asexually ✓   **D** Embryo transplantation ☐

2 Which of the following statements about the offspring of asexual reproduction are true? Tick the **two** correct options.

   A They show genetic variation. ☐

   B They are infertile. ☐

   C They are genetically identical to each other. ☐

   D They are genetically identical to the parent plant. ✓

   E They have stunted growth. ☐

   F They are prone to disease. ☐

Cloning Techniques

3 **a)** Which of the following techniques is suitable for cloning plants? Tick the **two** correct options.

   A Taking cuttings ☐   **B** Embryo transplantation ☐

   C Fusion cell ☐   **D** Tissue culture ✓

   E Computer-aided synthesis ☐

b) Which of the following techniques is suitable for cloning animals? Tick the **two** correct options.

   A Taking cuttings ☐   **B** Embryo transplantation ✓

   C Fusion cell ☐   **D** Tissue culture ☐

   E Computer-aided synthesis ☐

4 Are the following statements **true** or **false**? Write your answers in the spaces provided.

   a) A tissue culture technique produces offspring that are genetically identical to the parent plant.

   True

   b) Embryo transplantation produces offspring that are identical to its parents. _false_

   c) In adult cell cloning, the DNA from a donor animal is inserted into an empty egg cell. _True_

   d) Adult cell cloning produces offspring that are genetically identical to the parent animal.

   false

Genetics

Genetic Modification

1 Circle the correct options in the following sentences.

Genetic modification is a process in which genetic information from one **cell** / **organism** is transferred into another.

The genes are often transferred at a **early** / **late** stage of development, so that the organism will develop with the desired **behaviour** / **characteristics**

More organisms with the same characteristics can be produced if the genetically modified organism is then **cloned** / **reproduced**.

2 Give three reasons why genetic modification may be used in the production of food crops.

a) to improve resistance to pest or herbicides

b) to improve crop yield

c) to extend the shelf-life of fast-ripened crops

Insulin Production

3 Which organism is used to produce genetically modified insulin? Tick the correct option.

A Bacteria ✓   B Fungi ◻

C Insects ◻   D Viruses ◻

4 Which disease is the hormone insulin used to treat? Tick the correct option.

A Diabetes ✓   B Cystic Fibrosis ◻

C High blood pressure ◻   D Anaemia ◻

The Great Genetics Debate

5 Genetic engineering is a process that involves changing the genetic material of an organism. List three advantages of genetic engineering.

a) They have identified genes that control certian characteristics

b) They can determine whether a person's genes might increase the risk of them contracting a particular illness.

c) They soon might be able to remove faulty genes

6 Scientists have made great advances in their understanding of genes. However, some people are concerned that this knowledge may be misused. Suggest two ways in which this knowledge could potentially be misused.

a)

b)

The Theory of Evolution

1 When did life forms first exist on earth? Tick the correct option.

A 300 years ago ◯

B 3 million years ago ◯

C 3 billion years ago ◯

D We don't know ◯

2 What provides evidence for evolution? Tick the correct option.

A Animals ◯

B Fossils ◯

C Plants ◯

D Viruses ◯

Evolution by Natural Selection

3 Fill in the missing words to complete the following sentences.

Evolution is the change in a .... over many generations. It may result in formation

of a new ...., the members of which are better .... to their environment.

4 What is another name for evolution by natural selection? Tick the correct option.

A Creationism ◯

B Survival of the fittest ◯

C Genetic modification ◯

D Asexual reproduction ◯

5 a) If a dark coloured moth appears in a population of light coloured moths, what has happened?
Tick the correct option.

A A miracle has occurred ◯

B A genetic modification has occurred ◯

C A mistake has occurred ◯

D A mutation has occurred ◯

b) Under what circumstances might there be a change in the population of moths, where the number of
dark coloured moths becomes greater than the number of light coloured moths.
Tick the correct option.

A The dark moths eat the light coloured moths. ◯

B The dark moths are better adapted to the environment. ◯

C The male moths find the dark females more attractive. ◯

D The dark colouring is caused by an infectious disease. ◯

Extinction of Species

6 Give three factors that could contribute to the extinction of a species.

a) .... **b)** .... **c)** .... ◻

Pollution

The Population Explosion

1 Which two of the following effects have been created by an increase in the human population? Tick the **two** correct options.

    A Less land available for plants and animals ◯

    B Less waste being produced ◯

    C A reduction in pollution ◯

    D An increase in non-renewable energy sources being used ◯

2 The human population is increasing exponentially. What does **exponentially** mean?

Pollution

3 Which of the following gases pollute the air? Tick the **three** correct options.

    A Oxygen ◯   **B** Carbon dioxide ◯

    C Methane ◯   **D** Nitrogen ◯

    E Sulfur dioxide ◯   **F** Chlorine ◯

4 Which of the following are ways that smoke and waste gases from a power station can damage the environment? Tick the **three** correct options.

    A Cause acid rain ◯   **B** Decrease global warming ◯

    C Kill trees and plants ◯   **D** Increases carbon dioxide ◯

5 **a)** Name two pollutants that can affect the land.

    i)

    ii)

  b) Name two pollutants that can affect the water in our rivers and lakes.

    i)

    ii)

Indicators of Pollution

6 Give one example of a living organism that can be used as an indicator of pollution.

◯

Deforestation

1 What is deforestation? Tick the correct option.

A Planting new trees ◯

B Forest fires caused by hot weather ◯

C Cutting down large areas of forest ◯

D Polluting national parks with litter ◯

2 Give two reasons for large-scale deforestation.

a) ....

b) ....

3 What is the name of the gas given off when trees are burned? Tick the correct option.

A Carbon dioxide ◯

B Oxygen ◯

C Methane ◯

D Hydrogen ◯

4 Which biological process decreases the amount of carbon dioxide in the atmosphere?
Tick the correct option.

A Respiration ◯   B Exhalation ◯

C Photosynthesis ◯   D Deforestation ◯

5 a) Circle the correct option(s) in the sentences below.

When deforestation occurs in **tropical / arctic / desert** regions, it has a devastating impact on the environment.

The loss of **trees / animals / insects** means less photosynthesis takes place, so less **oxygen / nitrogen / carbon dioxide** is removed from the atmosphere.

It also leads to a reduction in **variation / biodiversity / mutation**, because some tree species may become **devolved / damaged / extinct** and **habitats / land / farms** are being destroyed.

b) What does **biodiversity** mean?

....

The Greenhouse Effect

The Greenhouse Effect

1 Fill in the missing words to complete the following sentences.

Some _____ in the atmosphere prevent _____ from escaping

into space. This is called the greenhouse effect.

The greenhouse effect is leading to global _____ .

2 Name two gases that contribute to the greenhouse effect.

a) _____

b) _____

3 Which of the following factors contribute to an increase in greenhouse gases?
Tick the **four** correct options.

A   Carbon offsetting ☐

B   Deforestation ☐

C   Burning fossil fuels ☐

D   Using renewable energy sources ☐

E   A growth in cattle farming ☐

F   Growing rice ☐

G   Forest management ☐

4 Which of the following are the negative effects of global warming? Tick the **two** correct options.

A   Climate change ☐

B   Erosion of buildings ☐

C   Deforestation ☐

D   A rise in sea levels ☐

E   An increase in available land ☐

F   Warmer summers ☐

5 Some people think that global warming does not affect them. Do you agree with them?
Explain your answer.

_____

_____

Sustainable Development

Sustainable Development

1 Fill in the missing words to complete the following sentences.

Sustainable development ensures that development can take place to help **improve / reduce / compromise** or sustain quality of life, without compromising the needs of future **space travel / generations / mutations**. It is an important consideration at local, regional and **sea / carbon dioxide / global** levels.

2 List the three key areas that sustainable development is concerned with.

a) ....

b) ....

c) ....

3 Briefly explain what sustainable resources are.

....

....

....

4 a) How can we maintain ocean fish stocks? Tick the correct option.

A Increase fishing ◯ **B** Increase fish farms ◯

C Introduce quotas ◯ **D** Stop eating fish ◯

b) Give one other method used to help maintain ocean fish stocks.

....

5 a) How can we maintain our forests and woodland? Tick the correct option.

A Have fewer national parks ◯

B Build more houses ◯

C Restock forests ◯

D Increase the burning of fossil fuels ◯

b) What is the key principle behind sustainable forest management?

....

c) Give two reasons why it is important to maintain our forests and woodland.

i) ....

ii) .... ▢

Elements and Compounds

Atoms and Elements

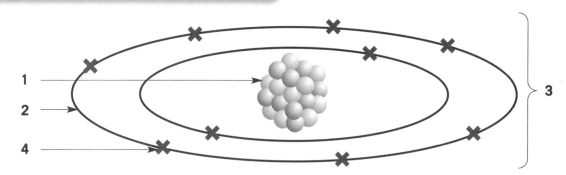

The diagram shows an atom and its subatomic particles.

Match statements **A, B, C** and **D**, with the labels **1–4** on the diagram. Enter the appropriate number in the boxes provided.

A Electron 4

B Electron shell 2

C Nucleus 1

D Atom 3

2 What is an element?

an element is a substance which only contains one sort of atom and are represented by different chemical symbols

3 Fill in the missing words to complete the following sentences:

a) The smallest particle to exist on its own is an _____ .

b) Electrons are negative particles found in the _electron shell_ of an atom.

c) The centre of an atom is called the _necleus_ .

d) In an element, all the atoms are _____ .

The Periodic Table

4 What are the chemical symbols for the following elements?

a) Sodium _Na_

b) Oxygen _O_

c) Iron _Fe_

d) Sulfur _S_

5 About how many elements are there in the periodic table? Tick the correct option.

A 80 ◯

B 90 ◯

C 100 ✓

D 150 ◯

Compounds

1 Draw lines between the boxes to match each key word to its definition.

Key Words

(1) Molecule

(2) Element

(3) Atom

(4) Compound

Definitions

(3) The simplest particle that can exist on its own.

(2) All of the atoms are the same in these substances, which are listed on the periodic table.

(4) A substance formed when two or more different types of atom are chemically joined together.

(1) A particle that consists of two or more atoms chemically joined.

2 Explain how electrons help atoms to join together.

....

....

Chemical Formulae

3 This diagram shows an ammonia molecule

Atoms can bond with each other to form molecules.

a) How many atoms are there in the molecule above? Tick the correct option.

   A 2 ◯   **B** 3 ◯   **C** 4 ✓   **D** 5 ◯

b) How many elements are there in the molecule above? Tick the correct option.

   A 1 ✓   **B** 2 ◯   **C** 3 ◯   **D** 4 ◯

c) What is the chemical formula of the compound shown? Tick the correct option.

   A $3NH$ ◯   **B** NH ◯   **C** N_3H ◯   **D** NH_3 ✓

Chemical Reactions

Chemical Reactions

1 **a)** Fill in the missing words to complete the following sentences.

In a chemical reaction, the substances that you start with are called _reactant_.

The new substances formed during the reaction are called _Product_.

b) Use the information in part a) to complete the following diagram, which represents a chemical reaction.

reactant	→	product

2 Equations are used to show what happens during a chemical reaction. Although the atoms are arranged differently, there is always the same number of atoms of each element on both sides of an equation.

Why is this?

because the same amount of have to be on each side for a balanced equation & no atoms are lost or made.

3 Sameena decided to investigate the thermal decomposition of calcium carbonate in a sealed conical flask. She measured the mass before and after the experiment.

a) What happened to the mass?

The mass will decrease

b) Explain your answer to part a).

because the calcium carbonate will decay

4 Why is it important for symbol equations to always be balanced? Tick the correct option.

A So that the mass of the reactants equals the mass of the products. ☐

B To show where all of the atoms have moved to in a chemical reaction. ☐

C Because no atoms are created or destroyed in a chemical reaction, just rearranged. ☑

D All of the above. ☐

5 Look at the equation below.

$$MgO + HCl \longrightarrow MgCl_2 + H_2O$$

Balance the equation, writing the correct version below.

$$MgO + HCl \longrightarrow$$

Writing Balanced Equations

1 When calcium carbonate is heated it undergoes thermal decomposition to produce calcium oxide and carbon dioxide.

a) Write a word equation for this reaction.

calcium carbonate $\xrightarrow{\text{heat}}$ calcium oxide + carbon dioxide

b) The formula for calcium carbonate is $CaCO_3$. Which of the following is the correct symbol equation for the thermal decomposition reaction? Tick the correct option.

A $CaCO_3 \longrightarrow 2CaO + CO_2$ ⬭

B $CaO \longrightarrow CaO + CO_2$ ⬭

C $CaCO_3 \longrightarrow CaO_2 + CO$ ⬭

D $CaCO_3 \longrightarrow CaO + CO_2$ ☑

2 Sulfuric acid (H_2SO_4) can react with calcium hydroxide ($Ca(OH)_2$) to form calcium sulfate ($CaSO_4$) and water (H_2O). Fill in the missing chemical symbols to balance the equation for this chemical reaction:

$$\underline{H}_2SO_4 + \underline{Ca}(OH)_2 \longrightarrow 2H_2\underline{O} + Ca\underline{S}O_4$$

3 Fill in the missing numbers to balance the equations below:

a) $CH\underline{} + 2O_2 \longrightarrow CO\underline{} + 2H_2O$

b) $N_2 + \underline{N}H_2 \longrightarrow \underline{2}NH_3$

4 Fill in the missing chemical symbols to balance the equations below:

a) $CuCO_3 \longrightarrow \underline{Cu}O + CO_2$

b) $2H\underline{Cl} + CuO \longrightarrow CuCl_2 + H_2O$

5 Oxygen only exists as a diatomic molecule, i.e. two oxygen atoms joined together.
Use the information above to write a balanced equation for the following reaction.
Copper (Cu) and oxygen (O) react to form copper oxide (Cu_2O).

$Cu + O \longrightarrow Cu_2O$

Limestone

Limestone (CaCO₃)

1. What type of rock is limestone? Tick the correct option.

   A Sedimentary ✓   **B** Metamorphic ☐

   C Igneous ☐   **D** Natural ☐

2. Limestone is made mainly from calcium carbonate, $CaCO_3$. How many atoms are in one molecule?

   3

3. **a)** What type of chemical reaction is used to turn limestone into quicklime? Tick the correct option.

   A Neutralisation ☐   **B** Electrolysis ☐

   C Oxidation ☐   **D** Thermal decomposition ✓

   b) What is the chemical name for quicklime?   Calcium Hydroxide

4. Calcium carbonate was heated in a Bunsen burner flame. In the reaction that took place, what is the name of the…

   a) reactant?   Limestone

   b) solid product?   Quicklime

   c) gas product?   Carbon Dioxide

5. **a)** Circle the correct option in the following sentence:

   Limestone can be heated to form calcium oxide. Water is then added to form calcium hydroxide. Calcium hydroxide is a strong **acid** / **alkali**.

   b) Why might a farmer want to add calcium hydroxide to their field?

   To neutralise any acidic soils

   c) What is another name for calcium hydroxide?   Slaked lime

6. Which of the following is a disadvantage of using limestone as a building material? Tick the correct option.

   A Limestone reacts with water. ☐   **B** Limestone reacts with acid in rain water. ✓

   C It is widely available. ☐   **D** It is easy to shape. ☐

Limestone (CaCO₃) cont.

1 Which of the following products are made from limestone? Tick the **four** correct answers.

A concrete ☑

B glass ☑

C granite ☐

D mortar ☑

E cement ☑

F marble ☐

2 **a)** Which limestone product would you use to make windows? _B - Glass_

b) Explain your choice of material for the window.

because it is see through

3 **a)** Which limestone product would you use to make bricks stick together? _Mortar_

b) Explain your choice of material for sticking bricks together.

because Mortar is liquid and when it is put between brick & hardens it sticks the bricks together

4 Tasmin made three different mixtures of concrete and tested them by hanging a load on them and recording the maximum force the concrete could take before it broke.

a) What is the dependent variable in this investigation? Tick the correct option.

A The amount of sand in the concrete. ☐

B The amount of cement in the concrete. ☐

C The load weight. ☐

D The amount of rock chippings in the concrete. ☐

b) What would be an appropriate unit for the dependant variable? Tick the correct option.

A Newton (N) ☑

B Kilogram (kg) ☐

C Litre (l) ☐

D Meter (m) ☐

c) What potential hazards should Tasmin be aware of in carrying out this practical investigation, and what precautions should she take?

Metals

Ores

1 (Circle) the correct options in the following sentences:

a) Unreactive metals that are found in the Earth's crust are said to be native. An example of a native metal is **gold** / **copper**.

b) Most metals occur as **nuggets** / **compounds** in the Earth's crust.

c) If the amount of the metal is sufficient for it to be economic to extract it from these rocks, they are also known as **minerals** / **ores**.

Extracting Metals from their Ores

2 Is the following sentence **true** or **false**?

Metals, like silver, occur naturally as pure metals because they are reactive. _____true_____

3 Which of the following metals can be found as pure elements in the Earth's crust?
Tick the **three** correct options.

A Pt ◯   B Ag ◯

C Zn ◯   D Fe ◯

E Au ◯

Iron

4 What type of chemical reaction happens in a blast furnace to produce pure iron from iron oxide?

A Neutralisation ◯   B Thermal decomposition ◯

C Reduction ◯   D Oxidation ◯

5 What is the difference between iron and steel? Tick the correct option.

A Iron is an element and steel is an alloy. ✓

B Iron is an alloy and steel is an element. ◯

C Iron is an element and steel is a compound. ◯

D Iron is a compound and steel is an alloy. ◯

6 Choose the correct word from the options given to complete the sentence below:

hard   **soft**   **brittle**   **shiny**

Pure iron is not used for construction because it is too _____soft_____ .

Alloys

1 Titanium is an expensive metal, but it has properties that make it very useful despite its price. Why is titanium often made into alloys?

A To make the metal harder ◯

B To make the metal cheaper ◯

C To make the metal stronger ✓

D All of above ◯

2 Choose the correct words from the options given to complete the following sentences.

alloy   **steel**   **element**   **atoms**

a) _____steel_____ is a mixture of iron, carbon and other metals.

b) Steel is an example of an _____alloy_____ .

c) Steel is harder than iron because it contains different sized _____atoms_____ , so the layers in its structure cannot slide over each other.

Steel

3 Sample 1   Sample 2

The two diagrams shown are of an element and an alloy. Match labels **A** and **B** with the correct diagram. Enter the appropriate number in the boxes provided.

A Iron ②

B Steel ①

4 Circle the correct options in the following sentences.

a) Low carbon steels are **resistant to corrosion / easily shaped**.

b) High carbon steels are **soft / hard**.

Smart Alloys

5 a) What is a **smart alloy**?

_____a metal that can retain it shape even though it has been bent_____

b) Why are new materials like smart alloys being developed?

_____

Metals

The Transition Metals

1 Fill in the missing words to complete the sentences below.

a) Transition metals are found in the _____ middle _____ of the periodic table.

b) They have all the usual properties of metals, which include being _____ conductors _____ of heat and electricity.

Extracting Transition Metals

2 What process is used to extract aluminium from its ore? _____

3 **a)** Titanium (Ti) is a very lightweight and strong metal.
Which of the following are uses for titanium? Tick the correct option(s).

A Aircraft

B Replacement joints

C Bikes

D Sunscreen

b) Aluminium is a good conductor of electricity, and is flexible and light.
Which of the following are uses for aluminium? Tick the correct option(s).

A Sandwich wrapping

B Water pipes

C Overhead power lines

D Drinks cans

Recycling Metals

4 Which of the following are advantages of recycling metals? Tick the two correct options.

A Uses less energy than extracting the metals from their ore. ✓

B Recycling facilities may be difficult to find.

C People are apathetic (can't be bothered).

D Less pollution is caused when metals are recycled compared to when they are extracted. ✓

5 Give two reasons why recycling reduces pressure on land compared to extracting more metal ore.

a) _____ because ore a found in the earth _____

b) _____

Crude Oil and Hydrocarbons

Crude Oil

1 What is a mixture? Tick the correct option.

    A More than one atom chemically joined ◯

    B More than one type of atom not chemically joined ◯

    C More than one type of substance not chemically joined ◯

    D More than one type of substance chemically joined ◯

2 What is crude oil? Tick the correct option.

    A A mixture of hydrocarbon molecules ☑

    B A mixture of carbon molecules ◯

    C A mixture of hydrocarbon atoms ◯

    D A mixture of hydrogen and carbon ◯

3 Is the following sentence **true** or **false**?

The longer the hydrocarbon chain in a molecule, the higher its boiling point. ....

Fractional Distillation

4 Explain how fractional distillation happens.

....

....

....

5 What do the fractions of crude oil contain? Tick the **three** correct options.

    A Mixture of compounds ◯   **B** Separate atoms ◯

    C Hydrocarbons ☑   **D** Alkanes ☑

    E Each fraction is a pure compound ◯

6 Circle the correct options in the following sentences:

Fractions with low boiling points come out at the **top** / **middle** / **bottom** of the fractionating column.

Fractions with high boiling points come out at the **top** / **middle** / **bottom** of the fractionating column.

7 Petrol has a relatively low boiling point. Is petrol a long or short-chain hydrocarbon?

.... ◻

Crude Oil and Hydrocarbons

Alkanes (Saturated Hydrocarbons)

1 Which elements are found in an alkane molecule?

2 (Circle) the correct option in the following sentence:

The atoms in alkanes are joined by **single** / **double** bonds.

3 What is an alkane? Tick the correct option.

   A An unsaturated hydrocarbon atom ⬭

   B An unsaturated hydrocarbon molecule ⬭

   C A saturated hydrocarbon atom ⬭

   D A saturated hydrocarbon molecule ⬭

4 **a)** What is the simplest alkane? _methane_

   b) How many hydrogen atoms does one molecule of the simplest alkane contain?

   4

   c) How many carbon atoms does one molecule of the simplest alkane contain?

5 What is the formula for propane? Tick the correct option.

   A C_2H_4 ⬭   **B** C_3H_6 ⬭

   C C_3H_8 ✓   **D** CH_4 ⬭

6 Fill in the missing words to complete the following sentence:

_____ hydrocarbons release _____ more quickly than

_____ hydrocarbons by burning, so there is greater demand for them as

_____ .

7 The chemical formula for one alkane is C_6H_{14}.
Draw the displayed formula for this alkane (i.e. showing the bonds between atoms using lines).

Burning Fuels

Revision Guide Reference: Page 45

1 What is a fuel?

_____

2 Which of the following is not a fuel? Tick the correct option.

   A Electricity ◯   **B** Hydrogen ◯

   C Coal ◯   **D** Petrol ◯

3 Which fuel produces only water as a waste product?

_____ Hydrogen _____

4 Fossil fuels contain mainly hydrocarbons. What products are produced when a hydrocarbon completely burns? Tick the **two** correct options.

   A Carbon ◯   **B** Carbon monoxide ◯

   C Carbon dioxide ☑   **D** Water ◯

5 Fill in the missing words to complete the sentences below.

   a) Many fossil fuels contain _____ as an impurity.

   b) As the fuels ____ combust ____ the impurity also reacts with oxygen and produces

      _____ .

   c) This gas can lead to _____ .

6 Choose the correct words from the options given to complete the following sentences.

   particles   **sulfur dioxide**   **carbon dioxide**   **water**

   a) The product of burning fossil fuels that can cause global warming is / are _ Carbon dioxide .

   b) The product of burning fossil fuels that can cause global dimming is / are _ particles _ .

7 Are the following sentences **true** or **false**?

   a) Acid rain can be reduced by removing sulfur dioxide before the waste gases of combustion are

      released to the atmosphere. _____

   b) Acid rain can be reduced by removing sulfur from fossil fuels before they are burned.

   _____

Cracking Hydrocarbons

Cracking Hydrocarbons

1 (Circle) the correct options in the following sentence:

Longer-chain hydrocarbons are **heated** / **melted** / **cooled** / **frozen** until vapour passes over a hot **plate** / **catalyst** / **area** / **shield**, producing short-chain hydrocarbons. This process is called cracking.

2 What type of reaction is cracking? Tick the correct option.

A Neutralisation ◯

B Electrolysis ◯

C Fractional distillation ◯

D Thermal decomposition ✓

3 Fill in the missing words to complete the sentence below:

Cracking hydrocarbons produces _____ and _____ .

Alkenes (Unsaturated Hydrocarbons)

4 (Circle) the correct option from the following:

Alkenes are **saturated** / **unsaturated** / **monomer** / **polymer** hydrocarbons.

5 Draw a representation of ethene in the box below.

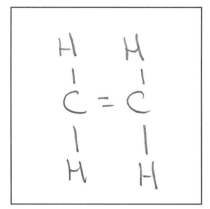

Making Alcohol from Ethene

6 a) Fill in the names of the two reactants in the spaces provided to complete the word equation to show how ethanol can be produced.

_____ + _____ → Ethanol

b) What is the name of the acid used as a catalyst in the reaction?

_____

Polymerisation

1 What is a monomer?

A monomer is an Alkene

2 What is the scientific word for a plastic? Tick the correct option.

A Alkene ◯   B Monomer ◯

C Polymer ◯   D Alkane ◯

3 Fill in the missing words to complete the sentences below:

Polymerisation is when ___monomers___ join together to form ___polymers___.

The compound changes from being unsaturated to ___saturated___.

4 What is the name of the polymer made from ethene molecules? Tick the correct option.

A Polymer ◯   B Plastic ◯

C Poly(ethene) ✓   D Poly(propene) ◯

5 What two factors do the properties of polymers depend on?

Representing Polymerisation

6 Complete the diagram, using displayed formulae, to show how ethene monomers are made into the polymer, poly(ethene).

Ethene Monomers (Unsaturated)	Poly(ethene) Polymers (Saturated)
H H H H | | | | C = C + C = C + | | | | H H H H	[H H] H H | | | | — C — C — C — C — | | | | H H H H

7 What is the general formula for polymerisation?

Polymers

Polymers

1 Give two examples of uses for a polymer.

a) ....

b) ....

2 What is a smart polymer?

....

3 Draw a line to link each polymer to its specific use.

Poly(ethene)	Protective packaging
Polystyrene	Crates and ropes
Poly(propene)	Plastic bags and bottles

Disposing of Plastics

4 What makes the disposal of plastics difficult? Tick the correct option.

A They are non-biodegradable ◯   **B** They are versatile ◯

C They are made from crude oil ◯   **D** They decompose easily ◯

5 What are the environmentally friendly ways to dispose of used plastics? Tick two correct options.

A Put them into landfill sites ◯   **B** Recycle ◯

C Burn them as fuel ◯   **D** Incinerate them ◯

6 Supermarkets have started giving their customers carrier bags made out of biodegradable plastic.

a) What does biodegradable mean?

....

....

b) What impact will using biodegradable plastic have on landfill sites?

....

Oils and Emulsions

Getting Oil from Plants

1 Fill in the missing words to complete the sentences below.

Oil can be extracted from plants by _____ them. In the oil extraction

process, water and impurities are removed by _____ .

2 Name three oils common in the food you eat.

a) _____   **b)** _____

c) _____

Vegetable Oils

3 Are the following statements **true** or **false**?

a) Vegetable oils provide a lot of energy.   _____

b) Vegetable oils provide a lot of water.   _____

c) Vegetable oils provide a lot of protein.   _____

d) Vegetable oils provide important nutrients.   _____

4 What two chemicals can be used to test for saturation? Tick the **two** correct options.

A Bromine ✓	**B** Iodine ☐
C Phosphorus ☐	**D** Chlorine ☐
E Hydrogen ☐	**F** Nickel ☐

Emulsions

5 Why don't oils dissolve in water?

_____

6 Which of the following statements are true? Tick the **two** correct options.

A Emulsions are thicker than oil or water.   ☐

B Emulsions have no coating ability.   ☐

C Emulsions are thinner than oil or water.   ☐

D Emulsions have a better texture and coating ability than oil or water.   ☐

Oils and Emulsions

Hydrogenation

1 a) Complete the following equation:

_____ + _____ → Saturated fat

b) What type of catalyst is used in this reaction?

_____

2 Give one reason why a food manufacturer might want to convert an unsaturated fat into a saturated fat.

_____

Additives

3 Give three reasons why E-numbers are added to our food.

a) _____

b) _____

c) _____

Chemical Analysis

4 The diagram below shows the results of a chromatography.

X is an ink sample. Identify whether the ink came from Pen **A, B, C, D** or **E** using the results below.

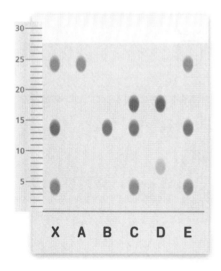

X = Pen _____

The Earth and Tectonic Theory

Structure of the Earth

1 The drawing shows the structure of the Earth.
Match statements **A**, **B**, **C** and **D** with the labels **1–4** on the diagram. Enter the appropriate number in the boxes provided.

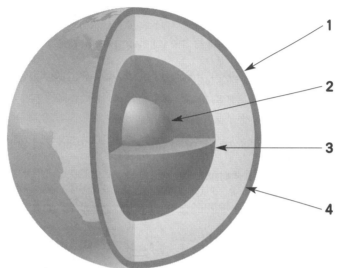

A Inner core   2

B Outer core   3

C Mantle   4

D Crust   1

2 What process is responsible for the rocks at the Earth's surface being continuously broken up, reformed and changed?

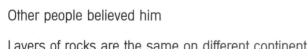 *continental Drift*

Tectonic Theory

3 a) Alfred Wegener was the scientist who first suggested the idea of plate tectonics. What evidence did Wegener use to support his theory? Tick the **three** correct options.

A The jigsaw fit of some continents ✔

B Other people believed him ☐

C Layers of rocks are the same on different continents ✔

D The periodic table ☐

E Fossil remains ✔

b) Alfred Wegener proposed that the movement of the Earth's crust was responsible for separating land masses. What name did he give to this process?

 *Continental Drift*

The Earth and Tectonic Theory

Tectonic Plates

1 What is the Earth's lithosphere? Tick the correct option.

A The core and mantle ◯   **B** The crust and upper mantle ◯

C The crust and atmosphere ◯   **D** The core, mantle and crust ◯

2 a) Fill in the missing words to complete the following sentence.

The Earth's lithosphere is cracked into a number of large pieces called

_____ _____ .

b) What causes the pieces of lithosphere to move? Tick the correct option.

A Convection currents in the crust ☑

B Convection currents in the mantle ◯

C Convection currents in the outer core ◯

D Convection currents in the inner core ◯

3 (Circle) the correct options in the following sentence:

Water / (intense heat) / gas / matter released by **(combustion) / currents / radioactive decay / oxygen** causes currents in the Earth's **(crust) / mantle / core / orbit**.

4 Use the words provided to fill in the missing words and complete the following sentences:

crust   **rock**   **slowly**   **radiation**   **down**   **convection**   **fast**

a) Hot molten _rock_ rises to the surface, creating new _crust_ .

b) The older, cooler crust then sinks _down_ where the _convection_ current starts to fall.

c) The land masses on these plates move _slowly_ .

5 Roughly how many major tectonic plates are there? Tick the correct option.

A 3 ◯   **B** 7 ◯

C 50 ◯   **D** 100 ◯

6 What natural disasters are common occurrences at plate boundaries?

Earthquakes & Volcanic eruptions

The Earth and Tectonic Theory

Tectonic Plate Movement

1 Describe the three ways in which tectonic plates can move in relation to each other.

a) ....

b) ....

c) ....

2 Fill in the missing words to complete the sentences below.

a) .... plate boundaries occur when plates collide and

one is forced under the other.

b) .... plate boundaries occur when plates move apart

and molten rock rises to the surface and forms new ocean floor.

c) How much new ocean floor is formed each year? Tick the correct option.

A A few millimetres ☑ **B** A few centimetres ☐

C A few metres ☐ **D** A few miles ☐

3 a) What natural disasters are common at plate boundaries? Tick the **three** correct options.

A Whirlwind ☐ **B** Hurricane ☐

C Volcanic eruption ☑ **D** Drought ☐

E Earthquakes ☑ **F** Tsunamis ☑

b) At what type of plate boundary are these natural disasters most likely to occur?

....

4 a) In your own words, describe how an earthquake occurs.

2 tectonic plates rub against each other

b) San Francisco experiences frequent earthquakes. Why do you think this is?

because they are on or near to tectonic plates

The Earth's Atmosphere

The Atmosphere

1 **a)** When the Earth was first formed, which gas did the atmosphere mainly consist of?
Tick the correct option.

A Oxygen ⬭   **B** Carbon dioxide ⬭

C Nitrogen ⬭   **D** Methane ⬭

b) Which gas does the Earth's atmosphere mainly consist of today? Tick the correct option.

A Oxygen ⬭   **B** Carbon dioxide ⬭

C Nitrogen ☑   **D** Methane ⬭

c) Over what time period did the change in the atmosphere take place?

A Hundreds of years ⬭   **B** Thousands of years ⬭

C Millions of years ⬭   **D** Billions of years ⬭

2 When the Earth first formed, there was intense volcanic activity.
Name three substances released by volcanic activity.

a) ....

b) ....

c) ....

3 How do most scientists believe the oceans were created? Tick the correct option.

A A chemical reaction between hydrogen and oxygen. ⬭

B The water vapour in the air boiled away. ⬭

C The water vapour in the air condensed. ⬭

D Plants respiring released water vapour. ⬭

4 Why did the evolution of green plants produce significant changes in the Earth's atmosphere?

....

....

5 Some carbon from the carbon dioxide in the air is taken out of the carbon cycle for long periods of time.
What happens to this carbon?

....

⬭

The Earth's Atmosphere

Composition of the Atmosphere

1 (Circle) the correct options in the following sentence.

a) For about the last **100 / 200 / 300 / 400** million years the proportion of gases in the atmosphere has been more or less the same.

b) There is about 80% **oxygen / nitrogen / argon / helium** and 20% **oxygen / nitrogen / argon / water** in the air.

c) There are also noble gases like **oxygen / nitrogen / carbon dioxide / helium**.

2 a) What distinguishes the noble gases from the other gases in the Earth's atmosphere? Tick the correct option.

A They are very volatile ☐   **B** They are unreactive ☐

C They are not naturally occurring ☐   **D** They are very dense ☐

b) Give one example of how a noble gas can be used.

helium

Changes to the Atmosphere

3 What are the two main reasons why the level of carbon dioxide in the atmosphere is increasing?

a) *deforestation*

b) *burning fossil fuels*

4 Fill in the missing words to complete the sentence.

The amount of carbon dioxide in the atmosphere is reduced by *photosynthesis* in green plants.

It is also removed through a reaction with _____ water, which produces carbonates.

Insoluble carbonates are deposited as _____, which forms rocks in the Earth's

_____ .

5 Is the increase in the level of carbon dioxide in the atmosphere good or bad? Explain your answer.

Conduction, Convection and Radiation

Conduction and Convection

1 Circle the correct options in the following sentences.

a) **Conduction** / **convection** is the transfer of heat energy through the movement of a substance.

b) **Conduction** / **convection** is the transfer of heat energy through a substance without the substance itself moving.

2 Fill in the missing words to complete the following sentences.

a) The process of heat energy being transferred along a metal bar is called _Conduction_ .

b) The process of heat energy being transferred through a liquid is called _Convection_ .

c) The process of heat energy being transferred through a gas is called _Convection_ .

3 Which of these statements about metals is **not** correct? Tick the correct option.

A Metals are good conductors because they contain free electrons that are able to move through the metal. ☐

B Metals are good conductors because they do not have free electrons that are able to move through the metal. ☑

C As the metal becomes hotter, its particles gain more kinetic energy. ☐

D As the metal becomes hotter, the free electrons move faster. ☐

4 a) Soup is cooked in a metal saucepan on a camp fire. How is the thermal energy transferred through the metal of the saucepan? Tick the correct option.

A By free electrons colliding with particles and other electrons. ☐

B By the atoms moving faster through the metal. ☐

C By infra red radiation moving through the metal. ☐

D By the heated metal expanding, becoming less dense and rising. ☐

b) How does the thermal energy spread through the soup? Tick the correct option.

A By hot air rising. ☐

B By the soup expanding, becoming less dense and rising. ☐

C By the soup contracting, becoming denser as it is heated. ☐

D By free electrons colliding with atoms. ☐

5 Circle the correct option in the following sentence.

Despite its name, a radiator heats the air in a room by **radiation / convection / conduction**.

☐

Conduction, Convection and Radiation

Radiation

Revision Guide Reference: Page 64

1 Fill in the missing words to complete the sentence below.

Infra red radiation is the transfer of heat energy by ___electromagnetic___ ___waves___.

2 How is the transfer of heat by infra red radiation different from the transfer of heat by conduction or convection?

___because no particles are involved___

3 a) Which of the following statements about the emission and absorption of infra red radiation are true? Tick the **two** correct options.

A Black is a good absorber and emitter. ☑

B Black is a poor absorber and emitter. ☐

C Black is a poor absorber and good emitter. ☐

D Black is a good absorber and poor emitter. ☐

E Light, shiny materials are good absorbers and emitters. ☐

F Light, shiny materials are poor absorbers and emitters. ☑

G Light, shiny materials are poor absorbers and good emitters. ☐

H Light, shiny materials are good absorbers and poor emitters. ☐

b) Which of the following statements about infra red radiation are true? Tick the three correct options.

A All objects emit and absorb thermal radiation. ☑

B The cooler the object, the more energy it radiates. ☐

C The amount of radiation an object gives out or takes in depends on its surface. ☐

D The amount of radiation an object gives out or takes in depends on its shape and dimensions. ☐

4 Circle the correct option in the following sentence.

An object will emit or absorb energy **faster** / **slower** if there's a big difference in temperature between it and its surroundings.

5 Samuel has a glass jug of very hot liquid. How would you suggest...

a) he can slow down the rate at which it emits heat to keep it hotter for longer?

___cover the jug in a light shiny material___

b) he can increase the rate at which it emits heat to cool it down quickly?

___cover the jug in a dark matt material___ ☐

Energy

Transferring and Transforming Energy

1 When a device transfers energy, only some of the energy is usefully transferred.

  a) What do we mean by **usefully transferred**?

    This means the energy has been rightfully transferred

  b) Fill in the missing words to complete the following sentences.

    The remaining energy is _transformed_ in a non-useful way. It is referred to as

    wasted energy.

2 Which one of the following statements is **not** true? Tick the correct option.

  A The energy transformed by a drill becomes increasingly spread out (dissipated). ☐

  B The energy transformed by a drill becomes difficult to use for further energy transformations. ☐

  C The energy transformed by a drill makes the surroundings a little warmer. ☐

  D The energy transformed by a drill is destroyed. ☑

Efficiency

3 A standard light bulb is not very efficient. Most of the electrical energy is transformed into what type

of energy? _heat_

4 **a)** If a device has an input of 100J/s, and 25J/s is transformed usefully to another form of energy, what is the device's efficiency? Tick the correct option.

    A 25% ☑   **B** 50% ☐

    C 75% ☐   **D** 4% ☐

  b) A fluorescent tube works at 50% efficiency. If it has 50J/s of electrical energy input, what will be its light output? Tick the correct option.

    A 25J/s ☑   **B** 100J/s ☐

    C 200J/s ☐   **D** 50J/s ☐

  c) A laptop can convert 400J of electrical energy to 240J of useful light and sound. What percentage of the energy is converted non-usefully (or wasted)? Tick the correct option.

    A 60% ☑   **B** 40% ☐

    C 24% ☐   **D** 50% ☐

Energy Transformation and Transfer

Energy Transformation

Revision Guide Reference: Page 66

1

| 1 | 2 | 3 | 4 |

These devices transfer electrical energy in different ways.

Match statements **A**, **B**, **C** and **D**, with the labels **1–4** on the diagram. Enter the appropriate number in the boxes provided.

A Movement (kinetic) energy ⟨3⟩   **B** Heat (thermal) energy ⟨1⟩

C Light energy ⟨2⟩   **D** Sound energy ⟨4⟩

Energy Transfer

2 a) Choose the correct words from the options given to complete the following sentences.

| decreased | National Grid | pylon | wattage |
| voltage | unaltered | distribution grid | increased |

The system of pylons that carry the generated electricity from the power station to our homes is

called the _National grid_. Transformers are used to change the _voltage_

of the current. If a step-up transformer is used, it is _increased_. If a step-down

transformer is used, it is _decreased_.

b) Circle the correct option in the following sentence.

Before the electrical energy reaches our homes its voltage needs to be **stepped up** / **stepped down.**

3 What type of electricity do we use in our homes? Tick the correct option.

A Direct current ◯   **B** Alternative current ◯

C Alternating current ☑   **D** AC/DC ◯

Energy Transformation and Transfer

Reducing Energy Loss

1 How can energy loss in power lines be reduced? Tick the correct option.

    A Increasing the current, decreasing the voltage ⬜

    B Increasing the voltage, decreasing the current ✓

    C Increasing current and increasing the voltage ⬜

    D Decreasing current and decreasing the voltage ⬜

2 Fill in the missing words to complete the sentence below.

For domestic use, electricity needs to be transmitted at a _____*high*_____ voltage and a

_____*low*_____ current.

Energy Calculations

3 a) What is the unit used to measure energy transfer from the mains? Tick the correct option.

    A kilowatt-hour, kWh ✓   **B** kilowatt, kW ⬜

    C joules per hour, J/h ⬜   **D** watts, W ⬜

b) What is the standard calculation used to find the total cost of energy transferred from the mains?

Total cost = number of kW/h × cost per kW

c) A 3kW electric fire is switched on for 2 hours. How much does it cost to use if a unit of electricity costs 5p? Tick the correct option.

    A 3p ⬜   **B** 30p ✓

    C 300p ⬜   **D** 1.2p ⬜

d) A 7kW electric shower is used for 30 minutes. What is the cost of using this if a unit of electrical energy costs 6p? Tick the correct option.

    A 21p ⬜   **B** 35p ⬜

    C 42p ⬜   **D** 1260p ✓

e) A 2kW hairdryer is used for 15 minutes every day. What is the total cost of using the hairdryer for a week if a unit of electricity costs 7p?

15 × 7 = 105   2 × 105 = 210kW/h   210 kW/h × 7p =

1470p   for 1 week.

Fuels

1 Fill in the missing word to complete the following sentence.

Fuels are substances which release useful amounts of energy when they are ___*burned*___.

Non-Renewable Energy Sources

2 What is the definition of a non-renewable energy source?

___*energy that is being used up faster than it is replaced*___

3 Is the following statement **true** or **false**?

Wood is a fossil fuel and is non-renewable. ___*false*___

4 a) Fill in the missing words to complete the following passage.

Nuclear fuel is used to generate electricity. A ___*chain reaction*___ is used to generate heat by

nuclear ___*fission*___. A heat exchanger transfers ___*thermal*___ energy from

the ___*reactor*___ to the ___*water*___. The water turns to

___*steam*___ and drives the ___*turbines*___.

b) Which of the following are nuclear fuels? Tick the **two** correct options.

A Coal ⬡   B Uranium ✓

C Oil ⬡   D Hydrogen ⬡

E Propane ⬡   F Plutonium ✓

c) Circle the correct options in the following sentences.

Nuclear fuels produce energy through a process called **fusion / (fission)**. Nuclear particles are released, which **fuse / (collide)** with the nuclei of other atoms, causing them to **grow / (split)** This causes a **(chain) / finite** reaction that generates huge amounts of energy.

5 Which of the following could not be used as fuel for a power station? Tick the correct option.

A Oil ⬡   B Gas ⬡

C Uranium ✓   D Water ⬡

Energy Sources

Non-Renewable Energy Sources

1 a) Below is a list of statements about coal as an energy source. Tick the statements that are disadvantages.

    A Cheap and easy to obtain ☐

    B Burning produces carbon dioxide ☑

    C Coal-fired power stations are quick to start up ☐

    D Sulfur in coal contributes to acid rain ☑

  b) Give one other disadvantage of using coal as a fuel.

    produces more CO_2 & SO_2 than oil or gas

2 a) Below is a list of statements about gas as an energy source. Tick the statements that are advantages.

    A Burning produces carbon dioxide ☐

    B Expensive pipelines are needed to transport it ☐

    C Sources are easy to locate ☑

    D Gas-fired power stations are the quickest to start up ☑

  b) Give one other advantage of using gas as a fuel.

    doesn't produce SO_2 (sulfur dioxide) so no production of acid rain

3 Which of the following is a valid argument against nuclear power stations? Tick the correct option.

    A They have high fuel costs. ☐

    B They produce gases that pollute the atmosphere. ☐

    C They have high decommissioning costs. ☑

    D They need to be in constant use to work efficiently. ☐

4 Which one of the following non-renewable energy sources does not produce carbon dioxide and sulfur dioxide? Tick the correct option.

    A Oil ☐   **B** Gas ☐

    C Nuclear ☑   **D** Coal ☐

5 Describe **two** advantages of using nuclear fuel instead of gas, coal and oil.

  a) There is less pollution

  b) Saves money for the long term

Renewable Energy Sources

1 What is the definition of a renewable energy source?

it won't run out and can continually be replaced

2 Many renewable energy sources are powered by the Sun or the Moon.

a) i) Fill in the missing word to complete the following sentence.

The gravitational pull of the moon creates _tides_ .

ii) Name one method of generating energy that relies on this.

Tidal (energy) Barrage

b) i) Fill in the missing word to complete the following sentence.

The Sun causes convection currents, which result in _wind_ .

ii) Name one method of generating energy that relies on this.

Nodding duck

3 Which type of power station involves the damming of a river flowing in an upland valley?
Tick the correct option.

A Nodding duck wave generator ◯   **B** Tidal barrage ◯

C Geothermal power station ◯   **D** Hydro-electric power station ☑

4 In some volcanic areas hot water and steam rise naturally to the surface of the Earth. The steam can be used to drive turbines. Which type of renewable energy source is this describing? Tick the correct option.

A Hydro-electric ◯   **B** Tidal ◯

C Wind ◯   **D** Geothermal ☑

5 a) Explain how nodding ducks are used to generate electrical energy.

is positioned in the sea so wave motion make the duck rock which is translated into a rotary movement which drives a generator.

b) Explain how a tidal barrage is used to generate electrical energy.

Tide comes in, water flows through a valve and becomes trapped, at low tide water is released through gap which has turbines which drives a generator

Energy Sources

Renewable Energy Sources

1 a) Draw lines between the boxes to match the energy sources with their advantages.

Source	Advantages
① Wind	② Fast start-up time
② Hydro-electric	④ Can produce electricity in remote location
③ Tidal	① Can be built offshore
④ Solar	⑤ Barrage water can be released when electricity demand is high

b) Draw lines between the boxes to match the energy sources with their disadvantages.

Source	Disadvantages
① Wind	③ A hazard to shipping
② Hydro-electric	④ Dependent on intensity of light
③ Tidal	① Causes noise and visual pollution
④ Solar	② Must be adequate rainfall in the region where the reservoir is

2 Which of the following statements is **not** true about a wave-driven power station? Tick the correct option.

A It can destroy habitats. ◯

B No pollutant gases are produced. ◯

C Variations of tides and waves affect output. ◯

D A lot of fuel is used. ◯

3 Which of the following statements is an advantage wind power has over nuclear power?
Tick the correct option.

A Wind power releases less CO_2 into the atmosphere. ◯

B Wind power releases less SO_2 into the atmosphere. ◯

C Wind turbines do not produce any dangerous waste. ◯

D Wind turbines take up less space when producing the same amount of electrical energy. ◯

Electromagnetic Radiation

1 Put the following four types of electromagnetic radiation in order of increasing wavelength. Write the numbers in the boxes provided (where 1 has the shortest wavelength and 4 has the longest wavelength).

A Infra red rays ◯   **B** Gamma waves ◯

C Radio waves ◯   **D** Ultraviolet rays ◯

2 Which of the following statements is correct? Tick the correct option.

A No electromagnetic radiation can be seen by the human eye. ◯

B All electromagnetic radiation is visible to the human eye. ◯

C All electromagnetic radiation can travel across a vacuum. ◯

D Only some of the parts of the electromagnetic spectrum can travel across a vacuum. ◯

3 Choose the correct words from the options given to complete the following sentences.

frequency	vacuum	medium	intensity	energy

a) All electromagnetic waves can travel across a .... because they do not need a

...., unlike sound waves.

b) All electromagnetic waves transmit ....

c) Each type of electromagnetic radiation differs from another in wavelength and

....

4 A wave has a wavelength of 1000m and a frequency of 300000Hz. What is its speed? Tick the correct option.

A 300m/s ◯   **B** 30000m/s ◯

C 300000m/s ◯   **D** 300000000m/s ◯

5 Circle the correct options in the following sentences.

When a wave is **reflected** / **refracted** / **absorbed** by a substance it makes the substance **move** / **hotter** / **colder** / **dissolve**.

It may create an **alternating** / **direct** / **modified** current of the same **wavelength** / **amplitude** / **frequency** as the radiation.

◯

Electromagnetic Radiation

Electromagnetic Radiation

1 The statements describe four uses of electromagnetic radiation. Match the types of electromagnetic wave **A, B, C** and **D** below with the statements **1–4**. Enter the appropriate number in the boxes provided.

1 Transmitting radio signals
2 Treating certain cancers
3 Transmitting satellite television channels
4 Security coding valuables

A Microwaves ◻ **B** Radio waves ◻

C X-rays ◻ **D** Ultraviolet rays ◻

2 Draw a line to match each part of the electromagnetic spectrum to the correct description.

Radio waves	Give images of broken bones, as they pass easily through soft tissue.
Microwaves	Kill cancerous cells and bacteria on food, but even exposure to low doses can cause cancer.
Infra red rays	Absorbed by the skin and felt as heat.
Ultraviolet rays	Used by mobile phone networks.
X-rays	Responsible for security coding.
Gamma rays	High levels of exposure for short periods can increase body temperature, causing tissue damage.

3 Which parts of the electromagnetic spectrum can be used for communications?

4 Microwaves are absorbed by the water particles in the food. What happens to the energy from the microwaves? Tick the correct option.

A It is emitted as light or sound energy. ◻

B It increases the potential energy of the water particles. ◻

C It increases the kinetic energy of the water particles. ◻

D It decreases the kinetic energy of the water particles. ◻

Electromagnetic Wave Communication

1 Explain briefly how electromagnetic waves carry sound as electrical signals.

Optical Fibres

2 a) Endoscopes are used by doctors to illuminate inside the ear. They make use of optical fibres. Explain why light stays inside the optical fibres.

b) Which of the following radiations is also used in fibre optic cables to transmit information. Tick the correct option.

A   Ultraviolet ( )   B   Microwaves ( )

C   Infra red ( )   D   Radio waves ( )

c) What are the advantages of using optical fibres over conventional copper electrical cables? Tick the **three** correct options.

A   They are lighter. ( )

B   They are thinner and take up less space. ( )

C   They cannot carry as many signals as copper cable. ( )

D   They are affected by electrical interference. ( )

E   There is energy loss along copper cables. ( )

Analogue and Digital Signals

3 Fill in the missing words to complete the sentences below.

A signal that continually varies its amplitude is an .... signal. These signals are more

prone to interference than digital signals because they can't ignore .... Digital

signals can be .... by computers unlike analogue signals. Digital signals only use

on and ....

Radiation

Isotopes

1 **a)** Fill in the labels A and B on the drawing of an atom.

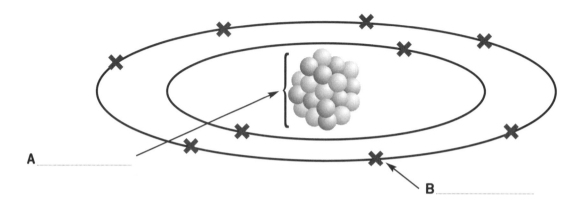

A _____

B _____

b) What is the nucleus of an atom made up of?

_____

Radiation

2 Circle the correct options in the following sentences.

a) A form of an element with the same number of protons but a different number of neutrons is called an **ion / isotope**.

b) When a radioactive particle collides with an atom, it can remove electrons and therefore create **ions / isotopes**.

c) A radioactive particle containing 2 protons and 2 neutrons is an **alpha / beta** particle.

3 **a)** What type of radiation consists of high energy electrons emitted from the nucleus?

_____

b) What type of radiation is not a particle but an electromagnetic wave? _____

4 Match statements **A, B, C** and **D** with the absorbers **1–4** below. Enter the appropriate number in the boxes provided.

1 Lead   **2** Thick concrete   **3** Paper   **4** A few cm of aluminium

A Needed to absorb gamma radiation. ☐

B All that is needed to absorb alpha radiation. ☐

C Needs to be several centimetres thick to absorb gamma radiation. ☐

D All that is needed to absorb beta radiation. ☐

Electric and Magnetic Fields

Revision Guide Reference: Page 80

1 **a)** Which types of radiation are deflected by electric fields and magnetic fields?

....

b) Why is this?

....

2 Which of the following statements about gamma radiation are true? Tick the correct options.

A It is not deflected by electric fields. ⬭

B It is deflected by magnetic fields. ⬭

C It is an electromagnetic wave. ⬭

D It travels faster than visible light in a vacuum. ⬭

Common Uses of Radiation

3 It is sometimes necessary to inject hospital patients with radioactive tracers. What is the purpose of injecting the tracer into the patient? Tick the correct option.

A To track the patient's movements around the hospital. ⬭

B To make the patient sleepy in preparation for an operation. ⬭

C To track the movement of substances around the patient's body. ⬭

D To find out the patient's blood group. ⬭

4 Give two common uses of gamma radiation.

a) ....

b) ....

5 Briefly describe how beta radiation is used in paper making.

....

....

....

Radiation

Acute Radiation Dangers

1 Fill in the missing words to complete the sentences below.

a) Radiation damage to cells in organs can cause _____. The larger the dose of

radiation, the _____ the risk.

b) The damaging effect of radiation depends on whether the source is _____ or

_____ the body.

2 a) Which type of radiation is not a problem to humans if the source is outside the body?
Tick the correct option.

A   Alpha   ◯   B   Beta   ◯

C   Gamma   ◯   D   Alpha and beta   ◯

b) Explain your answer to part a).

_____

c) Which type of radiation is most dangerous to humans when they are exposed to emissions from an external source?

_____

3 Although exposure to radiation carries a risk of cancer, alpha radiation is often used to treat mouth cancer. Explain why.

_____

_____

4 Which of the following are good precautions to take to protect a worker in a power station or factory from external exposure to radioactivity? Tick the **three** correct options.

A   Label all equipment with radioactivity stickers.   ◯

B   Avoid direct handling of the radioactive sources.   ◯

C   Site all power stations and factories using radioactive sources well away from towns.   ◯

D   Minimise exposure time and distance from the source of the radiation.   ◯

E   Use protective shields, screens and containers.   ◯

F   Do not carry instruments (e.g. a Geiger-Muller counter) to measure radiation levels.   ◯

Half-Life

1 Material from something that has once been alive can be dated by measuring its radioactivity. This is because all living things contain some radioactive carbon-14 atoms. Once a living thing dies it cannot absorb any more carbon-14. Carbon-14 decays by emitting beta particles and it has a half-life of 5730 years. This process is called carbon dating.

a) i) Which of the following could be dated by using carbon-14 measurements?
Tick the correct option.

A An iron horseshoe ⃝

B A leather belt ⃝

C A gold and silver piece of jewellery ⃝

D A Roman coin ⃝

ii) Explain your answer to part a).

....

....

b) A piece of charcoal from a stone age campfire contains 16 billion carbon-14 atoms. Approximately how many carbon-14 atoms would it have contained just under 6000 years ago?

....

c) What is a beta particle emitted from an atom of carbon-14? Tick the correct option.

A A decayed neutron. ⃝

B A helium nucleus. ⃝

C An electron from the inner part (the nucleus) of an atom. ⃝

D An orbiting electron from the outer part of an atom. ⃝

2 Why should a radioactive isotope have a short half-life if it is used as a tracer in someone's blood stream?

....

....

3 What does the term **half-life** mean?

....

⃝

The Universe

Observing the Universe

1 Match statements **A, B, C** and **D** with the advantages and disadvantages **1–4** below. Enter the appropriate number in the boxes provided.

1 Can be used 24 hours a day.

2 Cheaper to build, service and repair.

3 Can only be used at night and if the skies are clear.

4 Expensive to build, maintain and repair.

A An advantage of an optical telescope in space.

B An advantage of an optical telescope on Earth.

C A disadvantage of an optical telescope in space.

D A disadvantage of an optical telescope on Earth.

Telescopes

2 Draw lines to match each description with the correct type of telescope.

| Contains a parabolic glass mirror. |
| Orbits the Earth. |
| Contains two convex (converging) lenses. |
| Has a large (non-glass) parabolic dish and a receiver. |

| Radio |
| Reflecting |
| Refracting |
| Space |

3 Radio telescopes need to be very large compared to optical telescopes. Why is this?

Using Telescopes

4 Why are many telescopes placed on the top of mountains and in areas with low levels of pollution?

Red Shift

Revision Guide Reference: Page 83

1 a) Which of the following emit light, which provides possible evidence about the origin of the Universe? Tick the correct option.

A Planets ◯   **B** Moons ◯

C Comets ◯   **D** Galaxies ◯

b) The light emitted in part a) displays 'red shift'. What does this mean?

_____

_____

c) What do scientists suggest that red shift provides evidence of? Tick the correct option.

A The theory of evolution ◯

B Life elsewhere in the Universe ◯

C Global warming ◯

D The Big Bang theory ◯

2 Fill in the missing words to complete the sentences below.

Light that reaches the Earth from distant stars and galaxies is shifted _____ the red end of the electromagnetic spectrum. This is evidence that the galaxies are moving _____ from the Earth. This suggests that the Universe is _____.

3 The red shift effect is more exaggerated in the galaxies that are furthest away, What does this suggest?

_____

4 Which answer correctly describes the movement of the galaxies at the present time, **a)** relative to the Earth, **b)** relative to each other. Tick the correct option.

A a) moving away, **b)** moving closer ◯

B a) moving away, **b)** moving away ◯

C a) moving closer, **b)** moving closer ◯

D a) moving closer, **b)** moving away ◯

Periodic Table

Key

relative atomic mass
atomic symbol
name
atomic (proton) number

1	1
H	
hydrogen	
1	

1	2											3	4	5	6	7	0/8
																	4 **He** helium 2
7 **Li** lithium 3	9 **Be** beryllium 4											11 **B** boron 5	12 **C** carbon 6	14 **N** nitrogen 7	16 **O** oxygen 8	19 **F** fluorine 9	20 **Ne** neon 10
23 **Na** sodium 11	24 **Mg** magnesium 12											27 **Al** aluminium 13	28 **Si** silicon 14	31 **P** phosphorus 15	32 **S** sulfur 16	35.5 **Cl** chlorine 17	40 **Ar** argon 18
39 **K** potassium 19	40 **Ca** calcium 20	45 **Sc** scandium 21	48 **Ti** titanium 22	51 **V** vanadium 23	52 **Cr** chromium 24	55 **Mn** manganese 25	56 **Fe** iron 26	59 **Co** cobalt 27	59 **Ni** nickel 28	63.5 **Cu** copper 29	65 **Zn** zinc 30	70 **Ga** gallium 31	73 **Ge** germanium 32	75 **As** arsenic 33	79 **Se** selenium 34	80 **Br** bromine 35	84 **Kr** krypton 36
85 **Rb** rubidium 37	88 **Sr** strontium 38	89 **Y** yttrium 39	91 **Zr** zirconium 40	93 **Nb** niobium 41	96 **Mo** molybdenum 42	[98] **Tc** technetium 43	101 **Ru** ruthenium 44	103 **Rh** rhodium 45	106 **Pd** palladium 46	108 **Ag** silver 47	112 **Cd** cadmium 48	115 **In** indium 49	119 **Sn** tin 50	122 **Sb** antimony 51	128 **Te** tellurium 52	127 **I** iodine 53	131 **Xe** xenon 54
133 **Cs** caesium 55	137 **Ba** barium 56	139 **La*** lanthanum 57	178 **Hf** hafnium 72	181 **Ta** tantalum 73	184 **W** tungsten 74	186 **Re** rhenium 75	190 **Os** osmium 76	192 **Ir** iridium 77	195 **Pt** platinum 78	197 **Au** gold 79	201 **Hg** mercury 80	204 **Tl** thallium 81	207 **Pb** lead 82	209 **Bi** bismuth 83	[209] **Po** polonium 84	[210] **At** astatine 85	[222] **Rn** radon 86
[223] **Fr** francium 87	[226] **Ra** radium 88	[227] **Ac*** actinium 89	[261] **Rf** rutherfordium 104	[262] **Db** dubnium 105	[266] **Sg** seaborgium 106	[264] **Bh** bohrium 107	[277] **Hs** hassium 108	[268] **Mt** meitnerium 109	[271] **Ds** darmstadtium 110	[272] **Rg** roentgenium 111							

Elements which have atomic numbers 112–116 have been omitted.

*The lanthanides (atomic numbers 58–71) and the actinides (atomic numbers 90–103) have been omitted.

The relative atomic masses of copper and chlorine have not been rounded to the nearest whole number.

→ The lines of elements going across are called **periods**.

↓ The columns of elements going down are called **groups**.